SATURN

MARYSA STORM

BLACK RABBIT BOOKS

Bolt Jr. is published by Black Rabbit Books
P.O. Box 227, Mankato, Minnesota, 56002.
www.blackrabbitbooks.com

Michael Sellner, designer and photo researcher

Names: Storm, Marysa, author.
Title: Saturn / by Marysa Storm.
Description: Mankato, Minnesota : Black Rabbit Books, [2023]
Series: Bolt Jr. space explorer | Includes bibliographical references and index. | Audience: Ages 6-8 | Audience: Grades K-1 | Summary: "Suit up! Saturn is a planet like no other, and it's time to explore it through a colorful design that features vibrant photography, engaging infographics, and carefully leveled text"– Provided by publisher.
Identifiers: LCCN 2021022217 (print) | LCCN 2021022218 (ebook) | ISBN 9781623107772 (hardcover) | ISBN 9781644665985 (paperback) | ISBN 9781623107857 (ebook)
Subjects: LCSH: Saturn (Planet)–Juvenile literature.
Classification: LCC QB671 .S25 2023 (print) | LCC QB671 (ebook) | DDC 523.46–dc23
LC record available at https://lccn.loc.gov/2021022217
LC ebook record available at https://lccn.loc.gov/2021022218

Image Credits

123RF: bikejbarakus, 4–5; jgroup, 18–19; Alamy: dotted zebra, 12–13; Science Photo Library, 10–11; Dreamstime: Ratpack2, 20–21; Getty: Hulton Archive, 4–5; iStock: Elen11, 8–9, 22–23; Science Source: Detlev van Ravenswaay, 6–7, 22–23; MARK GARLICK, 1; NASA/Science Source, 16–17; Richard Bizley, 18–19; Shutterstock: Bikomins, 3, 24; Mopic, 14–15; NASA, 6–7; NASA images, 12–13; Vectorfair.com, 14-15; Vector Tradition, 10–11; Vadim Sadovski, Cover

Contents

Chapter 1
Meet the Planet 4

Chapter 2
Its Place in Space 10

Chapter 3
Explore! 16

More Information 22

CHAPTER 1

Meet the Planet

In 1610, an **astronomer** looked into the night sky. He studied Saturn. He became the first to discover its rings.

astronomer: an expert in the science of planets and stars

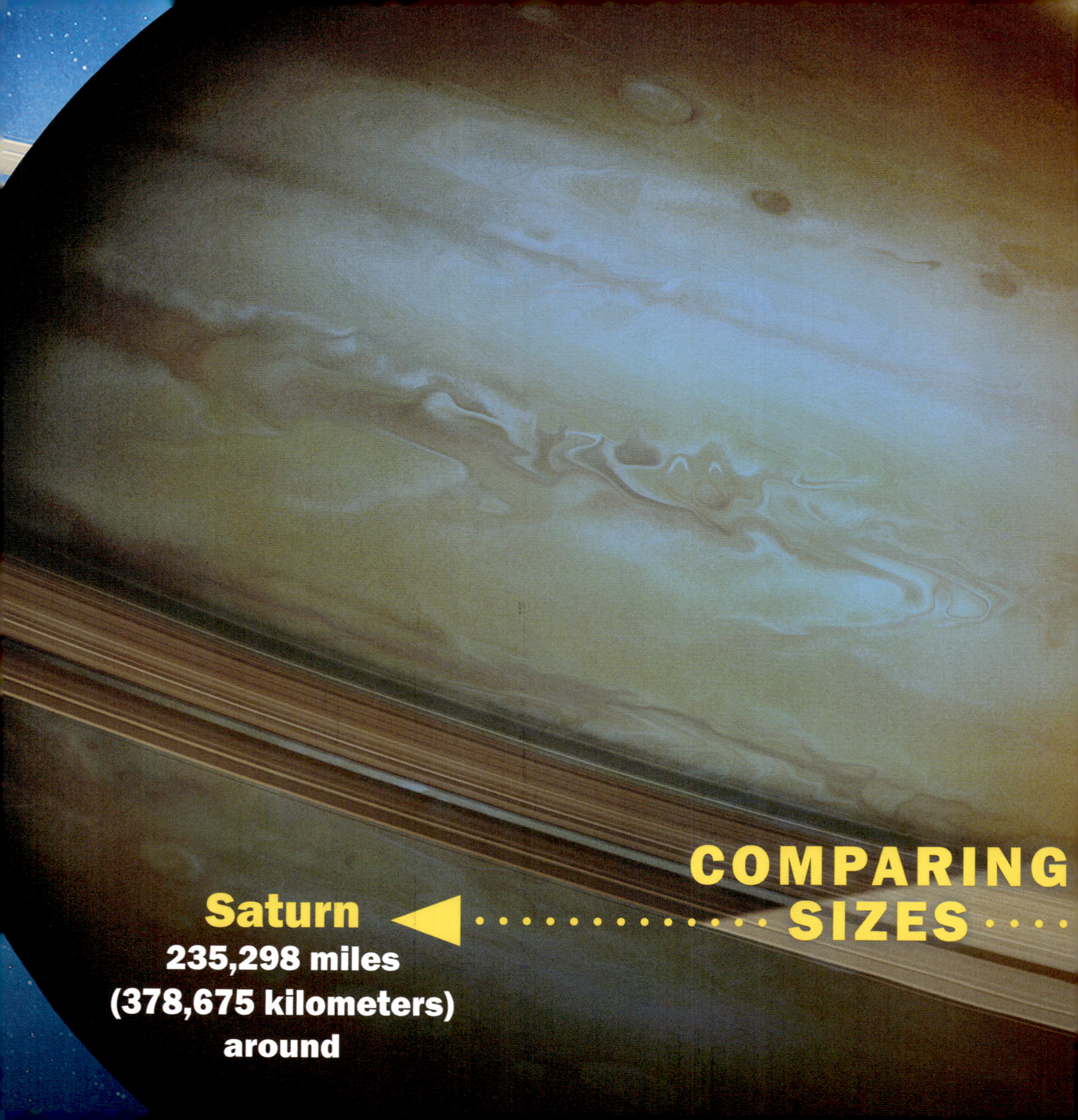

COMPARING SIZES

Saturn
235,298 miles
(378,675 kilometers)
around

A Big Planet

Saturn is known for its rings. It's also known for its size. It's about 10 times bigger than Earth.

Features

gold color
clouds
storms

CHAPTER 2

Its Place in Space

Saturn is an **outer planet**. It's the sixth planet from the sun. It's about 886 million miles (1.4 billion km) away.

outer planet: any of the solar system's planets that lie beyond the asteroid belt

FACT

Sunlight takes about 83 minutes to reach it.

Cold

Being so far from the sun makes Saturn cold. On average, it's -288 degrees Fahrenheit (-178 degrees Celsius). Life can't **survive** there.

survive: to remain alive

Saturn in the Solar System

Neptune
Uranus
Saturn
Jupiter
asteroid
belt

CHAPTER 3

Explore!

Four spacecraft have flown by Saturn. The last one to visit arrived in 2004. It was called *Cassini*. *Cassini* circled the planet for 13 years.

FACT

***Cassini* took about 450,000 pictures.**

The Future

People keep learning about Saturn. They study its moons too. Some could have life on them. Maybe someday, people will find out.

number of moons

at least 60

Bonus Facts

Its rings are made of rock and ice.

Some of its **moons** have oceans.

Saturn's days last about 10 hours.

It's made mostly of gas.

gas: a substance that is like air and has no fixed shape

READ MORE/WEBSITES

Finan, Catherine C. *Planets and Moons.* Space. Minneapolis: Bearport Publishing, 2022.

Loh-Hagan, Virginia. *Space.* Weird Science. Ann Arbor, MI: Cherry Lake Publishing Group, 2022.

Olson, Gillia M. *Curious about Saturn.* Curious About. Mankato, MN: Amicus/Amicus Ink, 2022.

Astronomy: Planet Saturn
www.ducksters.com/science/saturn.php

Mission to Saturn
kids.nationalgeographic.com/space/article/mission-to-saturn

GLOSSARY

astronomer (uh-STRON-uh-mer)—an expert in the science of planets and stars

gas (GAS)—a substance that is like air and has no fixed shape

outer planet (OU-ter PLAN-it)—any of the solar system's planets that lie beyond the asteroid belt

survive (sur-VIV)—to remain alive

INDEX

C

Cassini, 16

E

exploration, 16, 19

F

features, 4, 7, 8–9, 20, 21

M

moons, 19, 20

S

sizes, 6–7

T

temperatures, 13